Work 138

总是会有例外

There's always an Exception

Gunter Pauli

冈特·鲍利 著

凯瑟琳娜·巴赫 绘
高 青 译

学林出版社
www.xuelinpress.com

丛书编委会

主　任：贾　峰
副主任：何家振　闫世东　郑立明
委　员：牛玲娟　李原原　李曙东　李鹏辉　吴建民
　　　　彭　勇　冯　缨　靳增江

特别感谢以下热心人士对译稿润色工作的支持：

王必斗　王明远　王云斋　徐小帖　梅益凤　田荣义
乔　旭　张跃跃　王　征　厉　云　戴　虹　王　逊
李　璐　张兆旭　叶大伟　于　辉　李　雪　刘彦鑫
刘晋邑　乌　佳　潘　旭　白永喆　朱　廷　刘庭秀
朱　溪　魏辅文　唐亚飞　张海鹏　刘　在　张敬尧
邱俊松　程　超　孙鑫晶　朱　青　赵　锋　胡　玮
丁　蓓　张朝鑫　史　苗　陈来秀　冯　朴　何　明
郭昌奉　王　强　杨永玉　余　刚　姚志彬　兰　兵
廖　莹　张先斌

目录

总是会有例外	4
你知道吗？	22
想一想	26
自己动手！	27
学科知识	28
情感智慧	29
艺术	29
思维拓展	30
动手能力	30
故事灵感来自	31

Contents

There's always an Exception	4
Did you know?	22
Think about it	26
Do it yourself!	27
Academic Knowledge	28
Emotional Intelligence	29
The Arts	29
Systems: Making the Connections	30
Capacity to Implement	30
This fable is inspired by	31

穿山甲很是羡慕每天能在田野上开心地狂奔却时刻保持警觉的狐獴。

"你们一定是非洲这里最幸福的族群了。"穿山甲说道，"人们总是面带微笑地看着你们，因为你们似乎永远挂着笑脸，还有炯炯发亮的眼睛。"

A pangolin is admiring the meerkats. It just loves the way they run playfully across the field, and yet are very alert.

"You must be the happiest crowd here in Africa," claims the pangolin, "and when people watch you, they smile because you seem to smile all the time. Then you have these bright eyes that shine."

穿山甲很是羡慕狐獴……

A pangolin is admiring the meerkats ...

会认为我是来自另一个世界……

Would think I am from another world ...

"非常谢谢你!"狐獴妈妈回应着,并带点嘲讽口气说,"不过你看上去却很孤独,人们看你时,他们总是低着下巴。"

"我知道,如果人们看到我的舌头,他们会认为我是来自另一个世界。"

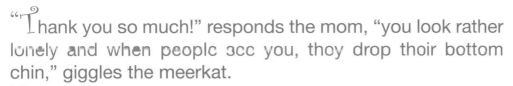

"Thank you so much!" responds the mom, "you look rather lonely and when people see you, they drop their bottom chin," giggles the meerkat.

"I know, and if the people would have a look at my tongue, they would think I am coming from another world."

"你需要用你长长的舌头,去伸进蚂蚁的巢穴和白蚁的土墩里。那你的舌头到底有多长呀?"

"我的舌头比我的身体还长,正是它让我的工作变得非常容易!"

"是吗?那么当你没在舔食土壤深处的昆虫时,舌头放在哪里呀?"

"You need that long tongue to get into the ants' nest and termites' mounds. How long is your tongue?"

"My tongue is longer than my body but it makes my work so easy!"

"What? And where do you put it when you are not licking insects deep in the ground?"

那你的舌头到底有多长呀?

How long is your tongue?

每年7000万只昆虫

70 million insects per year

10

"哦，我的舌头可以藏在我的胸膛里，因为我没有牙齿，所以我的身体里有足够的地方让我卷起舌头，并存放舌头粘住的东西。"

"明白了，蚂蚁和白蚁粘在你的舌头上。你吃过多少蚂蚁和白蚁了？"

"嗯，我现在仍然在发育成长中，估计当我长到和我妈妈一样大的时候，我每年能很轻松地吃掉7000万只昆虫。"

"Well, my tongue starts deep in my chest, but since I have no teeth there is enough place to roll up and store my sticky thing."

"I see, the ants and termites are stuck to your tongue. How many do you eat?"

"Well, I am still growing up, but when I will be as big as my mom, I could easily ingest 70 million insects per year."

"什么？我要是那样吃会消化不良的。"

"噢，不用担心，我的胃里有石头和刺，可以把吃进来的所有东西碾碎，变成一些又细又软的糊糊。"

"What? That would give me indigestion."

"Oh, do not worry. I have stones and spines in my stomach and all what comes in is crushed into a fine paste."

我的胃里有石头和刺！

Stones and spines in my stomach!

我的鳞片会保护我!

My scales protect me!

"你胃里有石头？那一定很重吧。是不是因为你肚子里有砖头，所以你才会总是滚来滚去而不是跑来跑去？"

"我身上的鳞片和你的爪子、犀牛角、头发，都是用一样的材料做的，但当我卷起来时，我还有个特别强大的翻转功能。我的鳞片盔甲占我体重的20%。它很坚硬，连狮子都嚼不动。"

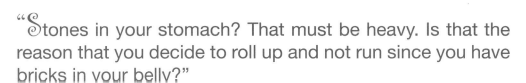

"Stones in your stomach? That must be heavy. Is that the reason that you decide to roll up and not run since you have bricks in your belly?"

"When I roll up, my scales made from the same as your claws, rhino horns and hair, but I have an extra strong version. My armour is 20 percent of my weight. It is strong, so strong that not even a lion can chew through it."

"这太令人肃然起敬了。"

"可不幸的是,我并没有受到尊重,我的朋友当中,有五分之一都死在了商店或餐馆里,因为人们认为吃了我的鳞片可以让他们更强壮、更健康。"

"That commands respect."
"Unfortunately, I am not respected, a fifth of my friends end up in shops or restaurants since people believe that eating my scales makes them stronger and healthier."

……我的鳞片可以让他们更强壮

... my scales will make them stronger

很可能会消失不在了……

May not be around much longer …

"那他们可以啃自己的指甲呀，它是由相同的材料做的。"狐獴笑着说。

"不幸的是，人们不知道吃我的鳞片是和啃指甲一样的。我们很可能会消失不在了，因为我们的脑袋和鳞片的价格飞涨，连我们最后的孩子们都因此受到威胁。"

"Well then they can better nibble on their fingernails, it is made from the same material," laughs the meerkat.
"Unfortunately, people do not know that eating my scales is the same a chewing on finger nails. We may very well not be around anymore since the last of our babies are threatened by these skyrocketing prices put on our heads and scales."

"这不可能是真的吧。我真希望你能够被保护起来。"

"很不幸啊!有那么多的动物和植物朋友被杀害了,我真希望我会是例外。"

"噢,你一定会有机会成为例外的。"

……这仅仅是开始!……

"It cannot be true. I hope that you will be saved."
"Unfortunately, there are so many animal and plants friends being killed, I hope that I will be the exception."
"Well, you certainly have a chance of being that exception."
... AND IT HAS ONLY JUST BEGUN!...

……这仅仅是开始！……

...AND IT HAS ONLY JUST BEGUN!...

Did You Know?
你知道吗?

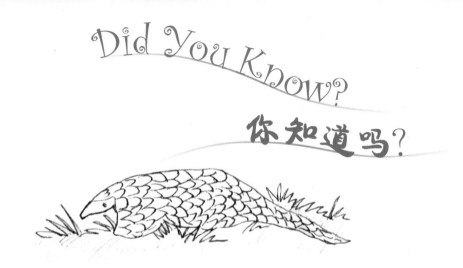

Pangolins are the most frequently trafficked mammals in the world. Over 100 000 are captured, traded and killed each year. The pangolin is considered by some as a health enhancing delicacy.

穿山甲是世界上被非法贩运得最厉害的哺乳动物。每年有超过10万只穿山甲被捕获、交易和杀害。有些人把穿山甲视为一种促进健康的美味。

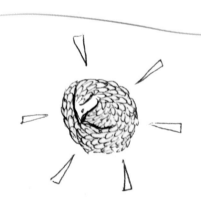

When the pangolin feels threatened it will first hide its head between its front legs, and if attacked it will roll into a ball using its very sharp scales for protection.

当穿山甲感到危险时,它会第一时间把头隐藏在它的前腿之间;当受到攻击时,它就会蜷缩成球状,并露出非常尖锐的鳞片来保护自己。

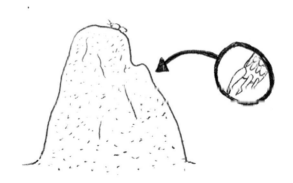

The name, "pangolin" is derived from the Malay word "pengguling", which translates to "something that rolls up". Pangolin feed on insects and use their claws to excavate termite nests.

穿山甲的名字来自马来语"pengguling", 相当于"卷起来某物"。穿山甲用它们的爪子挖掘白蚁巢穴, 并以吃昆虫为食。

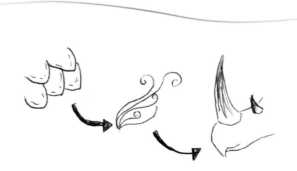

Pangolin scales are made of keratin, the same protein that makes up hair, nails, rhino horn, and the teeth of baleen whales. Their scales cover their whole body, except the underside which is covered in a few hairs.

穿山甲鳞片由角蛋白组成, 同样的蛋白质构成了毛发、指甲、犀牛角和须鲸的牙齿。鳞片覆盖了穿山甲全身, 除了由少量毛发覆盖着的下体。

Pangolin scales comprise 20% of their body weight. A pangolin can close its ears and nostrils to keep ants and termites stay while it is feeding.

穿山甲身上的鳞片占其体重的20%。穿山甲吃东西时可以选择关闭它的耳朵和鼻孔,好不让蚂蚁和白蚁跑掉。

One single pangolin consumes more than 70 million insects per year. It does not have any teeth, but its stomach contains small pebbles and keratin spines that grind their food.

一只穿山甲每年可以吞食超过7000万只昆虫。它没有任何牙齿,但胃里含有小鹅卵石和角蛋白刺用来磨碎食物。

Meerkats recognise each other's voices, they work together and even babysit each other's pups. Meerkats live in a matriarchal society, and take time to teach their young.

狐獴可以辨别出彼此的声音，它们一起工作生活，甚至照顾彼此的幼崽。狐獴生活在母系社会，并且花时间去抚育它们的幼崽们。

Meerkats are partly immune to the venom of scorpions. Nevertheless, the mother will not risk her young receiving a life-threatening sting while learning, through trial and error, to hunt scorpions. She will break off a scorpion's tail before feeding a live one to her young

狐獴对蝎子毒液有一定免疫力。但是母亲不会让孩子为了捕食蝎子而冒险去不断尝试这种威胁生命的"刺"。她会在给幼崽喂食活蝎子前先切断蝎子的尾巴。

Would you be able to eat if you have no teeth? And what about having teeth in your stomach instead of your mouth?

如果你没有牙齿，你能吃东西吗？ 如果牙长在胃里，而不是嘴里，你会怎么办？

Without pangolins, would there not be too many ants and termites?

如果没有穿山甲，就不会有太多的蚂蚁和白蚁了吗？

Does it make sense to eat keratin that's the same material as our fingernails?

吃角蛋白这种和我们的指甲一样的物质有意义吗？

Would you let your children play with scorpions?

你会同意让你的孩子玩蝎子吗？

How important are insect predators to an ecosystem? The pangolin eats 70 million ants a year, and if 100,000 pangolins are hunted illegally every year, that means that there are an extra 7,000,000,000,000 ants around to eat our harvest, or termites to eat away at the timber of our buildings. Make a list of all birds and mammals that are big insect eaters and that are being hunted or caught. Are these animals not important in the balance of the ecosystem? Will we have not an invasion of ants and termites, devouring our harvests and our homes should they become extinct? So build up mechanism for securing the survival of the most hunted animal in the world, because of the great service it is offering us, one that goes way beyond being just a delicacy.

动物昆虫的天敌有多重要？一只穿山甲一年吃7000万只蚂蚁，如果有10万只穿山甲被非法猎杀，这意味着将会有超过7万亿只蚂蚁侵吞我们的劳动果实，或者白蚁蛀毁我们的家园。列出所有能大量吞食昆虫但现在正在被捕猎的鸟类和哺乳动物。这些动物在维持生态系统的平衡中难道不重要吗？万一这些物种灭绝了，我们不是将面临蚂蚁和白蚁的入侵，看着它们吞噬我们的劳动成果或家园吗？所以要建立机制，保护那些被猎杀得最厉害的动物，因为它们可为人类提供除了美味以外的更多价值。

TEACHER AND PARENT GUIDE

学科知识
Academic Knowledge

生物学	狐獴属于獴科；狐獴是狐獴属的唯一成员；狐獴社会化极强，形成族群，成员间互相帮助；真菌类以角蛋白为食。
化 学	角蛋白是一种纤维状蛋白质；角蛋白和甲壳素有区别。
物 理	当穿山甲快速疾走时，它用后腿行走，并用尾巴来保持稳定和平衡。
工程学	角蛋白能阻止环境污染物和病毒侵入身体；法医学家可以从微小的毛发或指甲样品中判断营养物质、药物和毒物的存在；毛发中含有角蛋白。
经济学	市场每公斤3000美元的价格需求是推动穿山甲非法交易的动力，这种不断提升价格的不良非法行为，刺激了更多的非法猎杀活动。
伦理学	穿山甲（和许多其他动物）在被囚禁时压力很大，会过早死亡；动物非法贸易加快了穿山甲灭绝的速度。
历 史	穿山甲已存在约8000万年了；穿山甲从生活在白垩纪晚期的小型原始胎生哺乳动物进化而来。
地 理	狐獴生活在南非的卡拉哈里沙漠，以及纳米比亚、博茨瓦纳和安哥拉等地的干旱地区和热带稀树草原。
数 学	计算质量和物理强度。
生活方式	动物贩卖被接受是因为我们相信这些动物有益于我们的健康；必须容忍规则存在例外。
社会学	母系社会与父系社会的对比；明星力量的重要性：大象、老虎和犀牛都非常受欢迎，但大量濒临灭绝的穿山甲却不为人所知。
心理学	狐獴看上去有一种永恒的微笑，它会传染，让人感觉很好，便于建立和谐关系。
系统论	穿山甲在抑制昆虫数量上起到了关键作用，如果这种抑制作用被削弱，就需要扩大使用化学品。

教师与家长指南

情感智慧
Emotional Intelligence

穿山甲

穿山甲喜欢狐獴的微笑。微笑是会传染的,穿山甲提到人们通常是以微笑回应微笑的。穿山甲也意识到人们看到它们时的惊讶表情。穿山甲不介意谈论它奇特的特性,包括它能吃掉数百万只昆虫的胃口以及它胃里有牙齿的事实。穿山甲感谢狐獴对它如此大的消化功能的赞赏,并马上解释了它的胃和鳞甲的更多细节。穿山甲很有自我意识,而且完全掌握了自己的功能。穿山甲意识到它的独特性可能危及它的生存。实际上,穿山甲非常担心它的生存状况,并希望自己会是个例外。

狐獴

狐獴感激穿山甲对它的微笑的积极评价,这是其整个家族的标志。狐獴很同情地指出穿山甲孤独时流露出的一丝丝嘲笑,也证实了它与其他动物们相比是那么不同。狐獴很自信地问及一些私人问题,比如舌头和饮食风格。狐獴还关注到穿山甲的进食方式会不会可能导致消化问题。狐獴正在寻找更好的理解,并准备推测穿山甲的习性,而当它完全了解了穿山甲的所有功能,狐獴表示出了对穿山甲的敬慕之情。狐獴很务实,以玩笑的口吻讲述其化学成分(角蛋白)的进化,让穿山甲担忧其生存的紧张心情轻松下来,并鼓励了穿山甲。

艺术
The Arts

微笑是会传染的。就像通常可以看到的熟悉的笑脸符号,我们现在有微笑似乎印在脸上的狐獴。让我们画一下素描。观察狐獴一系列照片,它的脸部轮廓就是一个笑脸。只用笔来勾画脸、鼻子、眼睛和嘴的线条。重点是,把你画完的作品拿给别人看,观察人们是如何反应的,他们也在微笑吗?如果你画的笑脸狐獴引发人们的微笑,那么你画了一幅优秀的画。如果人们没有微笑……那你的画就要继续改进。

TEACHER AND PARENT GUIDE

思维拓展
Systems: Making the Connections

　　传统医学对人们的健康非常重要。这种知识已被代代相传，印度的阿育吠陀医学甚至流传了几千年。虽然现代医学确实成功地治愈了许多疾病，但它更多的只是关注治疗，且会产生副作用。传统医学也有副作用，其中一个不幸的负面影响是对某些植物和动物的市场需求在快速增长，而这些物种的繁殖力不够快就会跟不上市场需求。穿山甲的减少，高涨的价格，刺激了更多的非法贸易，这导致该物种濒临灭绝。《濒危野生动植物种国际贸易公约》将因日益增长的非法贸易需求而导致濒临灭绝的物种列了清单。穿山甲是世界上非法交易最为严重的哺乳动物。许多专家质疑，如果从现代科学、基础化学和可持续性的角度看，牺牲这种哺乳动物的理由是否合乎逻辑。人们最感兴趣的是穿山甲的鳞片。这些鳞片由蛋白质构成，称为角蛋白。从化学上讲，它与我们指甲的材料一样。所以很难理解利用这种动物的必要性，因为它是一种明显没有任何特别之处的东西。这是保护动物的关键，并让我们更好地理解需求增加导致脆弱生命的灭绝。然而很明显，仅靠压制手段打击非法贸易作用不大。也许有效的办法是科学分析这方面的实例，让人类更好更广泛地了解濒危物种的独特性。这需要经常重新审视类似现实，让偷猎者在不同的更公正的经济发展中寻找机会从而放弃偷猎。而关于如何创造一个灭绝规则例外的新兴愿景，更容易在一个积极的气氛中实现。微笑无疑增加了使科学和社会融为一体的氛围。希望这能确保通过医学进行个体健康探索，以证实地球上各种物种健康发展不是相互排斥的，而是能够共享生存的。

动手能力
Capacity to Implement

　　我们如何能对一个几乎不了解的奇怪动物产生注意呢？在研究穿山甲，了解这类哺乳动物独特之处时，我们应该把穿山甲和大象、犀牛、老虎的图片挂在一起吗？或者应该展示一下穿山甲婴儿如何坐在母亲的尾巴上？还是应该展示被捕获的死亡的2000只穿山甲的惊悚照片？虽然人们很容易为这些被杀死的哺乳动物及其濒临灭绝的状况感到伤心困扰，但更应该思考有没有另外一种拯救这种动物免于灭绝的方法。所以，你的任务就是要抛弃所有消极情绪，专注于用积极的方式去调动人们用微笑拯救世界。

教师与家长指南

故事灵感来自

阮文泰
Thai Van Nguyen

阮文泰拥有理学硕士学位（环境科学），并获得澳大利亚国立大学的环境管理与发展硕士学位，还有肯特大学和德雷尔野生生物保护信托基金会颁发的德雷尔濒危物种管理的研究生证书。阮文泰在祖国为保护野生动物而奋斗近10年，2014年发起成立了"拯救越南野生动物"组织。他在2005年加入菊芳国家公园（越南）的亚洲穿山甲保护方案。他被选为40个野生动物保护英雄之一。他正与国家警察和当地社区合作，为营救和释放穿山甲而努力。他与国际自然保护联盟合作，确保穿山甲得到国际上的关注。

图书在版编目（CIP）数据

总是会有例外：汉英对照/（比）冈特·鲍利著；（哥伦）凯瑟琳娜·巴赫绘；高青译. ——上海：学林出版社，2017.10
（冈特生态童书. 第四辑）
ISBN 978-7-5486-1262-9

Ⅰ. ①总… Ⅱ. ①冈… ②凯… ③高… Ⅲ. ①生态环境－环境保护－儿童读物－汉、英 Ⅳ. ① X171.1-49

中国版本图书馆 CIP 数据核字（2017）第 143530 号

© 2017 Gunter Pauli
著作权合同登记号　图字 09-2017-532 号

冈特生态童书
总是会有例外

作　　者——	冈特·鲍利
译　　者——	高　青
策　　划——	匡志强　张　蓉
责任编辑——	胡雅君
装帧设计——	魏　来
出　　版——	上海世纪出版股份有限公司 学林出版社
	地　址：上海钦州南路 81 号　电话/传真：021-64515005
	网　址：www.xuelinpress.com
发　　行——	上海世纪出版股份有限公司发行中心
	（上海福建中路 193 号　网址：www.ewen.co）
印　　刷——	上海丽佳制版印刷有限公司
开　　本——	710×1020　1/16
印　　张——	2
字　　数——	5 万
版　　次——	2017 年 10 月第 1 版
	2017 年 10 月第 1 次印刷
书　　号——	ISBN 978-7-5486-1262-9/G.488
定　　价——	10.00 元

（如发生印刷、装订质量问题，读者可向工厂调换）